Learning about Simple M

This book shows how you can teach about simple machines using everyday materials.
Each activity follows a four-step scheme:

1. Set the stage.
2. Ask a question.
3. Confirm the principle.
4. Discuss applications.

There are two facing pages for each of the simple machines covered in this book. One page gives instruction for the teacher describing how to conduct the activity on the playground. The other page is a reproducible response or activity page for students.

Simple Machines

1. inclined planes
2. levers
3. gears
4. pulleys

Materials You'll Need

- a small table such as a card table
- 2 three-gallon (11 liter) buckets (with handles)
- a coffee cup or measuring cup
- 2 wood planks, 4' (1.22 m) and 8' (2.44 m)
- a sturdy box or block about 1' (30.5 cm) high
- a skateboard
- 25' (7.6 m) piece of clothesline cord
- 1 ten-speed or 5-speed bike
- up to 4 small pulleys

Plus tools and devices such as:

axe	pliers	crowbar
wedge	wheelbarrow	scissors
screw	nutcracker	shovel
bolt	paper cutter	hoe
wrench	broom	fishing pole
screwdriver	tweezers	hammer

Activity Options

There are several ways that these activities can be organized, depending on your schedule, the amount of available materials, etc. Look through all of the activities in this book before starting. This will help you decide how you will want to proceed. Here are some possibilities:

- Do the playground activities one at a time, followed by the corresponding classroom follow-ups.

- Do the playground activities all at once, then do the classroom follow-ups.

- Do all of the playground activities first, then do the follow-ups with small groups rotating through six centers.

Note: A cup is not a machine. It is used in this activity as a simple device to begin student exploration of simple machines.

Playground Activity #1

What is a Simple Machine?

Set the Stage

Take your students out to the sandbox on the playground where you have set up a table containing an empty bucket and a cup. Place a second bucket full of sand on the ground. Ask a single student to lift the bucket full of sand and pour some of it into the bucket on the table. Watch this step carefully to see that the child does not strain himself/herself. Establish that the bucket is too heavy to lift safely.

Ask the Question

Pose this question: "Is there an easy way that one person can get this sand onto the table?" Have students talk over the question. Giving them as little help as possible, guide them to the conclusion that one person could ladle the sand up into the bucket one cup at a time.

Pick a student to ladle the sand up and have the group count the number of times the cup has to be filled and lifted to get the job done. After the sand has been moved, ask the group to decide whether it was more or less work to do the job this way.

Through discussion, develop the idea that it's about the same amount of total work whether you lift a heavy load all at once or a light load in many smaller amounts. It takes less force the second way, but you have to do it more times. Devices that make jobs easier like this are called "simple machines."

Simple Machines and Mechanical Advantage

Explain that machines are devices that make work easier, but that there is always a trade-off. In this case the cup acted as a "machine" by making the lifting lighter. The trade-off was having to make more lifts. A device that makes the job five times easier, but that takes five times as long is said to have a "mechanical advantage" of five. Have students fill in the answers to the questions on their activity sheet (page 3). Have students pair off and practice explaining the concepts of simple machines and mechanical advantage to each other. Then, review the answers on the activity sheet as another way of reviewing and clarifying the concepts.

Applications of the Principles

Explain that what was learned during this activity is the basic principle for all simple machines. Remind them to look for that principle in all of the activities to follow, by continually asking themselves these questions:

1. How is this machine making the job easier?
2. What is the trade-off?
3. What is the "mechanical advantage"?

What is a Simple Machine?
Student Sheet

You saw that lifting a heavy bucket of sand
was too difficult for one person to do alone.
How was the job made easier?

_____

_____

Did the job take less total work to do that way?

☐ yes   ☐ no

In what way was the job easier?

_____

_____

In what way was the job harder?

_____

_____

Using a device to reduce the amount of force needed to do a job can make the job take longer, but it can be worth the longer time. Sometimes there is no other way you could get the job done. What do you call devices that help in this way?

_____

What do you call the help a device provides in terms of needing less force or effort?

_____

Draw a picture to show that
lifting one heavy container is
equal to lifting five light
ones:

=

The "mechanical advantage" of using these smaller containers would be _____

Inclined Planes

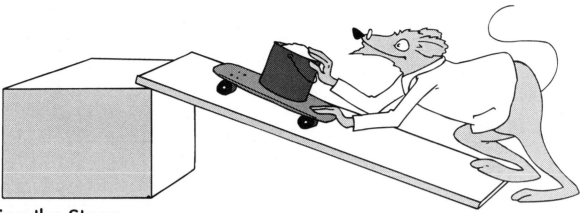

Setting the Stage

Place a box about 1 foot (30.5 cm) high (an inverted milk carton container works well) out on the playground. Set these items next to the box: a 4' (1.22 m) plank and an 8' (2.44 m) plank, a skateboard, and a big bucket of sand. This time there is no cup available for ladling the sand.

Asking the Question

How can I get all of the sand up on the box at one time? Once your students think of using the planks and skateboard to roll the sand up onto the box, ask more questions such as: "Since the skateboard added to the weight of the bucket and sand, was it worth using it?" "Which plank should we use? Why?" "Would a longer plank be even better? How long would be too long? Why?"

Confirming the Principle: Inclined Plane

A tilted board or ramp is a simple machine called an "inclined plane." Review the meaning of simple machines and mechanical advantage. Then ask questions such as: "What was easier about using an inclined plane than just lifting the bucket?" "What was the trade-off?" (less weight to lift but more distance to travel).

To determine the mechanical advantage of an inclined plane, compare the height to be lifted to the length of the inclined plane:

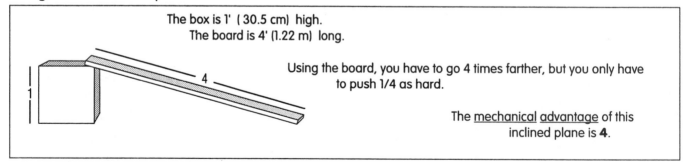

The box is 1' (30.5 cm) high.
The board is 4' (1.22 m) long.

Using the board, you have to go 4 times farther, but you only have to push 1/4 as hard.

The <u>mechanical</u> <u>advantage</u> of this inclined plane is **4**.

Applications

Ask your students where they have seen ramps. Can they think of other uses for ramps? Have someone describe what a mountain road is like. Does it go straight up the mountain? Why not? How about hiking trails? What is a "switchback" on a hiking trail? What determines how many switchbacks need to be carved out of a hillside?

Name _____

Date _____

Inclined Planes
Student Sheet

You saw in the demonstration how to move a heavy bucket up to the top of a box.
Use what you learned to answer these questions.

1. How were boards used to make the job of lifting heavy sand easier?

_____

_____

2. Did the job take less total work to do that way?

☐ yes   ☐ no

3. In what way was the job easier?

_____

_____

4. In what way was the job harder?

_____

_____

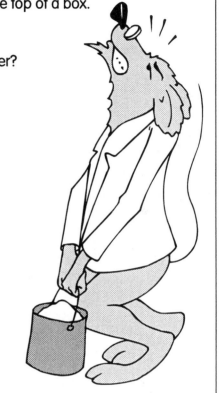

What is the mechanical advantage of each of these inclined planes?

mechanical advantage = _____

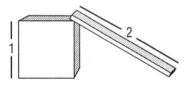

mechanical advantage = _____

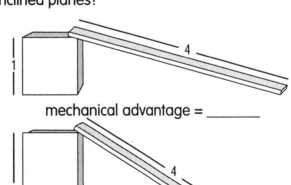

mechanical advantage = _____

mechanical advantage = _____

Name some examples of where inclined planes are used:

_____

_____

Other Kinds of Inclined Planes

Setting the Stage

Lay out several items on your table or out on the playground. Include a shovel, a hoe, an axe, a wood-splitting wedge, a screw, a bolt, a wrench, a crowbar, a screwdriver, a hammer, etc.

Asking the Question

Have your students name each item. See if they can tell you how each is used. Ask them to decide how many of these devices are examples of a simple machine. Confirm or explain that all of the tools involve at least one principle of a simple machine.

Now ask students to decide how many of these tools are examples of an inclined plane. It may take a while to convince students that all of the tools they have seen involve an inclined plane. This will not be obvious at first, as the inclined planes are being used in different ways than the one you demonstrated previously.

The inclined planes here are of two basic types:

Wedge

A wedge is an inclined plane that forces things apart as it is pushed between them.

Screw

A screw is an inclined plane that goes around and around so that it will push its way into or through a hard material as it is turned.

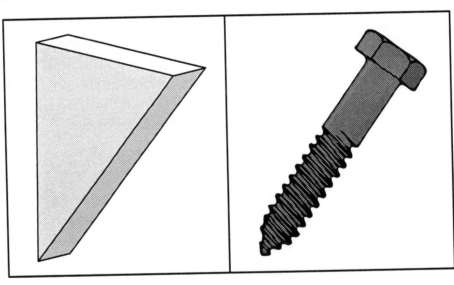

Confirming the Principle

Inclined planes aren't used just to raise heavy things to a higher level. They can also be used to push things apart (a wedge) or to hold things together (a screw).

Applications

Ask your students to think of some ways the different types of inclined planes are used.

The construction industry is filled with uses for all three kinds of inclined planes. Tilted boards are used to push materials up to the construction site. Wedges are used to separate things, and screws are used to put things together

Name _____

Date _____

Other Kinds of Inclined Planes
Student Sheet

The first type of inclined planes you worked with were used to move heavy things up to a higher level. There are two other uses for inclined planes.

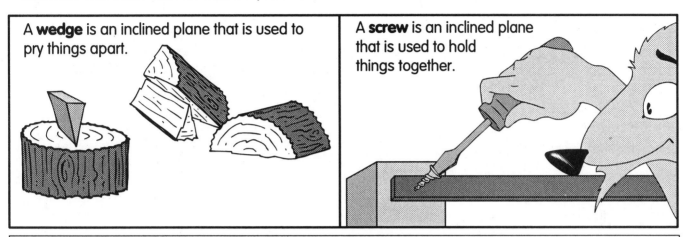

A **wedge** is an inclined plane that is used to pry things apart.

A **screw** is an inclined plane that is used to hold things together.

Here are some commonly used tools. Look for wedges or screws in these tools. They all have one or the other. Draw an arrow to the part of each tool that works as either a wedge or a screw and label it.

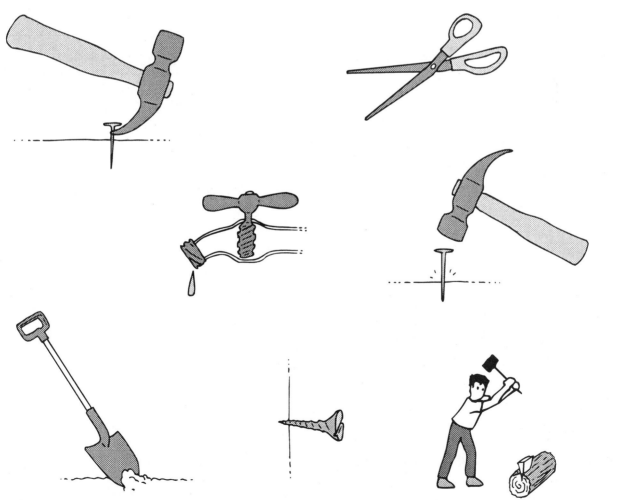

Levers

Setting the Stage

You will need access to a sandbox with a firm border or an alternative such as a low, but sturdy box. This is to serve as a fulcrum (balance point) in the following activity. Set a heavy bucket of sand next to the sandbox border (or box). Have the planks from activity #2 nearby. Tell your students the following story.

Once upon a time there was a king who offered the hand of his daughter in marriage to the prince who most impressed him when the prince came to the castle. However, the king had blocked the forest-lined road to the castle with a huge boulder. Prince #1 tunneled under the boulder to reach the castle, but the king was not impressed. Prince #2 built a ladder and climbed over the boulder, but the king was still not impressed. Prince #3 decided to move the boulder completely out of the road, so that others could easily pass by. The king was very impressed by this. But...the question is, how did the prince move such a giant boulder?

Asking the Question

Have your students think of the bucket of sand as the huge boulder. Have them try to think of a way the prince could have moved the boulder by himself. When they think of using a plank as a lever braced against the border of the sandbox or the box that is there, verify that this is what the prince did. If your students have difficulty coming up with this solution, be ready to guide them in that direction by asking questions.

Confirming the Principle: Lever

The simple machine the prince used was a lever. Extend the activity by laying one of your boards across the edge of your sandbox to act as a fulcrum. Show your students how the weight of the bucket of sand seems to change, depending on where the fulcrum is placed. The closer the weight is to the fulcrum, the lighter it seems to be. Let the students feel the weight change themselves.

How do you determine the mechanical advantage of this type of lever? With a lever, it is the distance from the fulcrum to the weight divided into the distance from the fulcrum to the person doing the pushing.

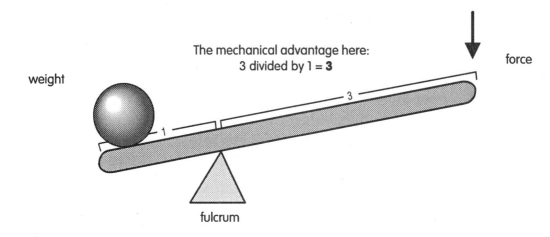

weight

The mechanical advantage here:
3 divided by 1 = **3**

force

fulcrum

Applications

Discuss various devices that work on the lever principle, including:

- crowbar
- tire iron
- paint can opener

Simple M

Inclined Planes

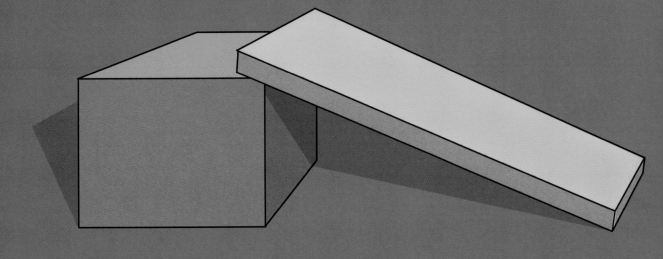

wedge

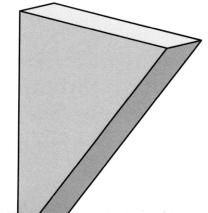

screw

Can you spot the simple machines at this construction site?

- inclined plane
- wedge
- screw
- 1st class lever

- 2nd class lever
- 3rd class lever
- pulley
- gears

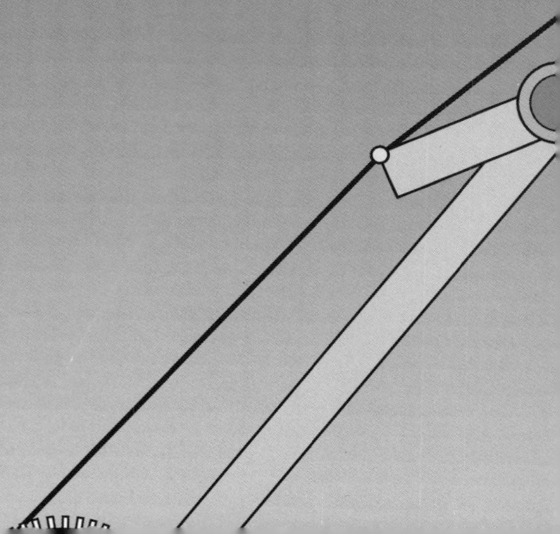

Simple M

Machines

Gears

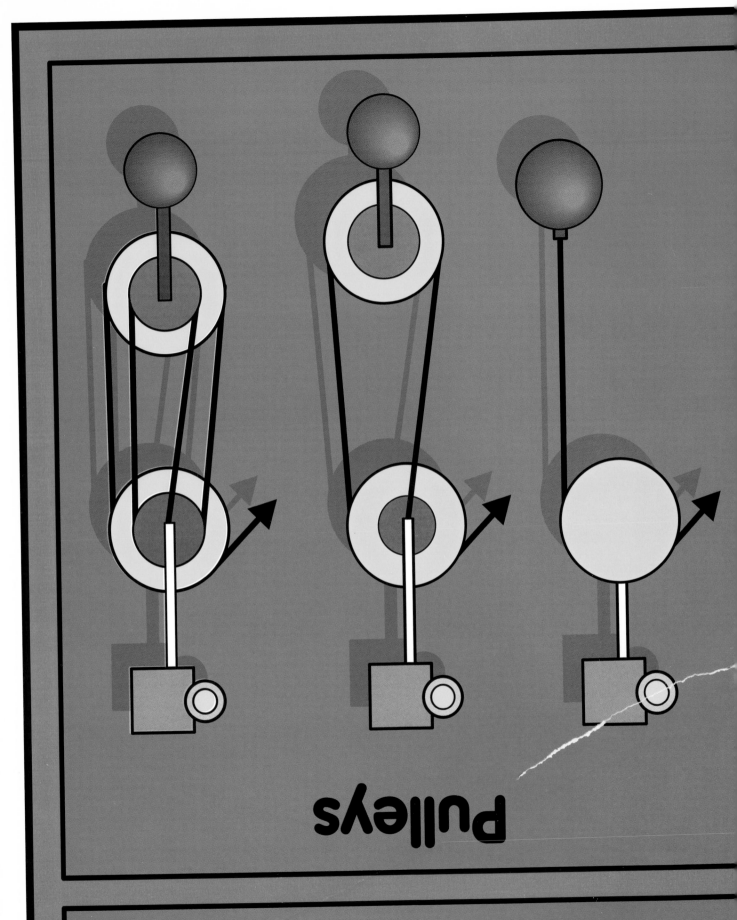

Pulleys

Playground Physics · EMC841

Levers

first class

second class

third class

Name _____

Date _____

Levers

You saw how a lever was used to lift the bucket.
Use what you learned to help you answer these questions.

How was a board used to make the job of lifting a heavy weight easier?

_____

_____

Did the job take less total work to do that way?

☐ yes   ☐ no

In what way was the job easier?

_____

_____

In what way was the job harder?

_____

_____

What is the mechanical advantage of each of these levers?

mechanical advantage = _____

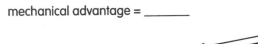
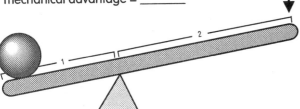

mechanical advantage = _____

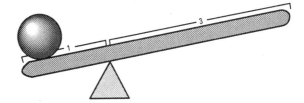

mechanical advantage = _____

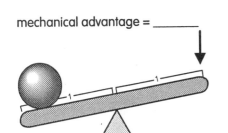

mechanical advantage = _____

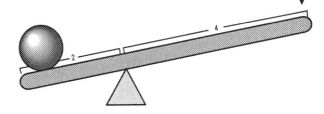

Name some examples of how levers like these are used:

_____ _____

What do you call the balance point of a lever? _____

9   Playground Physics

Other Kinds of Levers

There are three kinds of levers. Every kind of lever has:
- • a balance point (fulcrum)
- • a point where the weight is
- • a point where force is applied

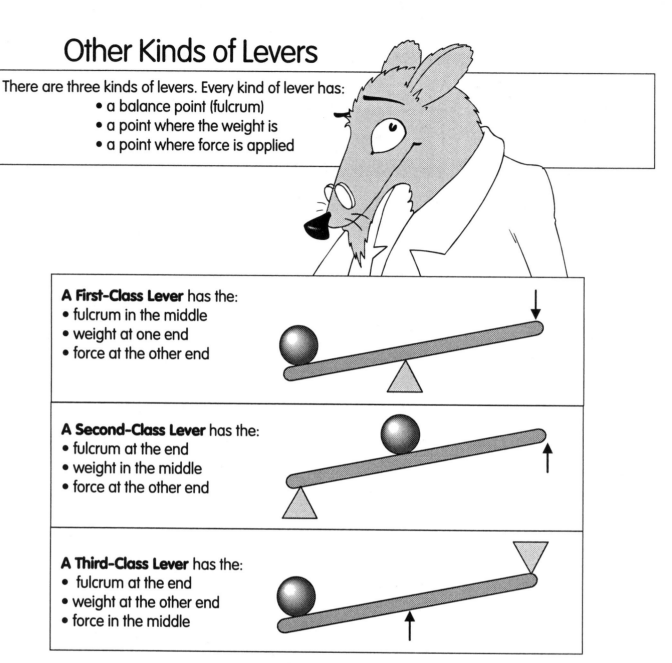

A First-Class Lever has the:
- • fulcrum in the middle
- • weight at one end
- • force at the other end

A Second-Class Lever has the:
- • fulcrum at the end
- • weight in the middle
- • force at the other end

A Third-Class Lever has the:
- • fulcrum at the end
- • weight at the other end
- • force in the middle

Draw each of the three classes of levers on the chalkboard for your students to see. Explain the parts of each type. Then demonstrate the devices listed below in random order. Help your students decide to which class of lever each belongs.

First-class levers	Second-class levers	Third-class levers
crowbar	wheelbarrow	shovel
scissors	nutcracker	broom
pliers	paper cutter	tweezers

Point out that when using third-class levers, the positioning of the force will determine whether the advantage is power or speed. If you put your hand near the dirt when lifting a shovel, you can lift a heavy load. If you put your hand further up the handle, you can't lift as much dirt, but you can dig faster. You have speed instead of power. You can also throw the dirt farther. Your students can get the feel of that idea really well by changing their hand position on a shovel or a broom. As they move hand positions, they are changing the mechanical advantage.

Have your students practice what they have learned using the sorting activity on page 11.

 Playground Physics

Other Kinds of Levers

You have learned that there are three kinds of levers. You have seen examples of each kind. The pictures below are of types of levers. Cut the pictures apart and sort them into three groups according to the type of lever.

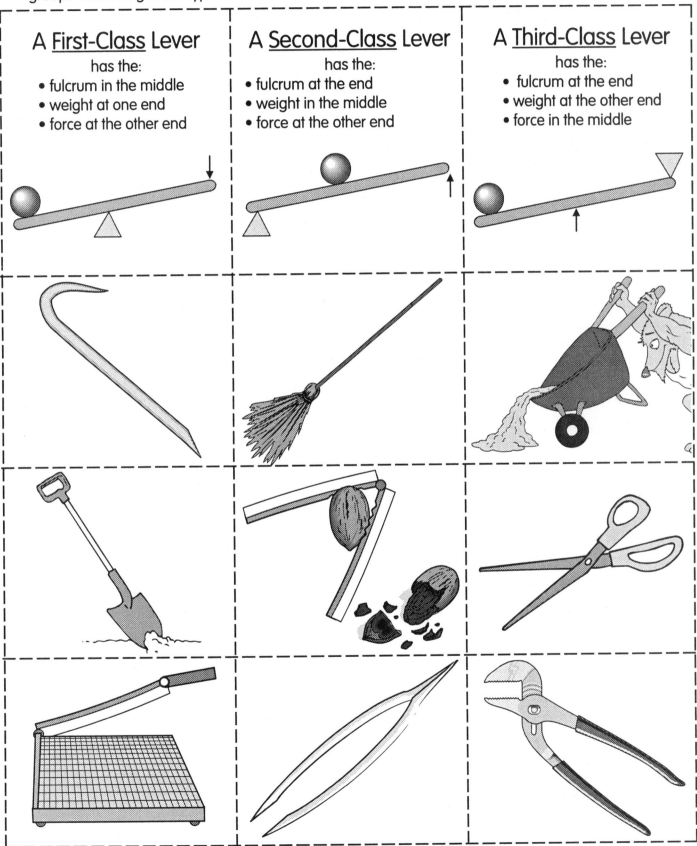

A <u>First-Class</u> Lever
has the:
- fulcrum in the middle
- weight at one end
- force at the other end

A <u>Second-Class</u> Lever
has the:
- fulcrum at the end
- weight in the middle
- force at the other end

A <u>Third-Class</u> Lever
has the:
- fulcrum at the end
- weight at the other end
- force in the middle

Playground Physics

Gears

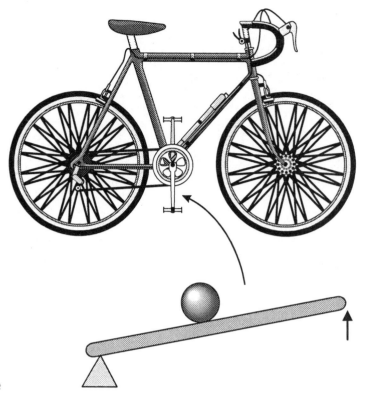

Setting the Stage

The activities you have done with levers serve as a natural lead-in to gears. Take your students out to the playground and show them a ten-speed bike. Ask them to try and find a lever on the bike. If they need help, draw their attention to the pedal mechanism. Discuss how the length of the rod leading to the pedal makes a big difference in how hard it can be to pedal the bike. The rod length changes the mechanical advantage.

Asking the Questions

Ask your students if they can see another kind of simple machine changing the mechanical advantage while a bike is being pedaled. Ride the bike around the playground following a predetermined circle, changing the gears as you go. See if your students can see a difference in how you are pedaling at different speeds and in different gears.

Ask "When is it an advantage to use a low gear?" (when your legs are going around quickly but the bike isn't going very fast) "When is it an advantage to use a high gear?" (when you are pedaling very slowly, but it is hard to push the pedals around)

Have your students count the number of times your feet must go around to make one trip around the circle in first gear. Then have them count the number of turns of the pedal it takes to go around the circle in tenth gear. Ask them what the difference is in mechanical advantage between first and tenth gear.

Confirming the Principle: Gears

Some gears push against each other directly, such as the gears in a car transmission. On the bike, the gears are connected by a chain. You can figure out the mechanical advantage of gears by comparing the number of teeth on the gears that drive each other. That comparison tells you the mechanical advantage.

If the smaller gear pushes the larger one, it has to go around faster, but it is easier to push. On a bike, that's good for going up a hill.

If the bigger gear pushes the smaller one, it's harder to push, but one push will make you go farther. That's good when you are on the flat ground and moving fast.

Applications

Gears are very common in vehicles and lifting devices such as cranes. Once your students understand how gears work, they will begin to see them everywhere they go.

       Playground Physics

Gears
Student Sheet

Now that you have seen how gears can give a bike rider a mechanical advantage, let's see if you understand the basic principle of gears.

Here are two gears. The big one has 16 teeth. The little one has 8. In both examples below, you would be turning the wheel on the left with the crank handle to make the one on the right go around.

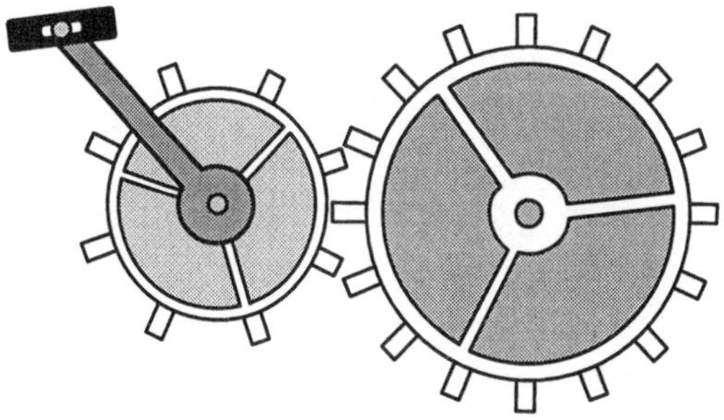

This big wheel on the right is really heavy. You'd have a tough time getting it to turn.

When the little wheel goes around once, how many times will the big one go around? _____

Will the little wheel go slower or faster than the big wheel?

_____

Will adding the small wheel make it easier to turn the big one?

_____

How could you make it even easier to turn the big heavy wheel?

_____

_____

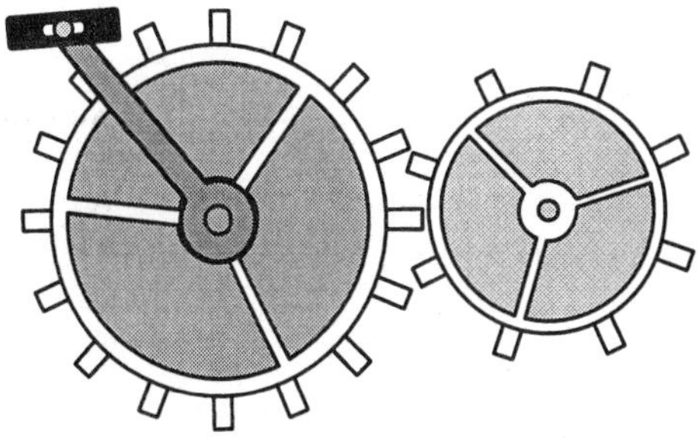

You need to get the little wheel on the right to go really fast. It is attached to a grinding wheel.

When the big wheel goes around once, how many times will the small one go around? _____

Will the little wheel go slower or faster than the big wheel?

_____

Will adding the big wheel make it easier to turn the small one?

_____

How could you change this arrangement to make the small wheel go even faster?

_____

_____

13

Pulleys

Setting the Stage

Set this activity up beneath the monkey bars. You will need your heavy bucket of sand and a table containing some small pulleys and 25' (7.6 meters) of clothesline cord. Because of the friction in pulleys, you may only be able to use half a bucket of sand or even less for this activity.

Asking the Question

Once again ask "How can I lift this heavy bucket of sand?" Your students will notice the pulleys right away. The question is, how can you use them in such a way that there is a mechanical advantage? The best way to pursue the answer is to have students make suggestions and try them out. Look at how these arrangements come out:

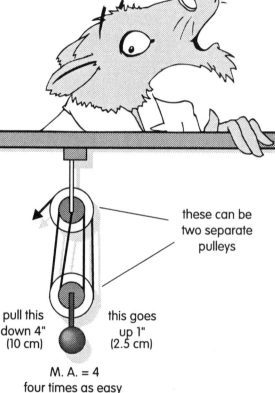

these can be two separate pulleys

pull this down 4" (10 cm)

this goes up 4" (10 cm)

M. A. = 1
it won't take less force with this pulley, in fact it will take more because of friction

pull this down 4" (10 cm)

this goes up 2" (5 cm)

M. A. = 2
twice as easy to pull it up, but only goes half as far

pull this down 4" (10 cm)

this goes up 1" (2.5 cm)

M. A. = 4
four times as easy to pull it up, but only goes one-fourth as far

Confirming the Principle: Pulleys

The way to determine the mechanical advantage of a pulley system is to count the number of strands of line that are supporting the weight. Each of those lines have to be shortened to pull up the weight. If you pull out four inches of line and the weight only goes up one inch, the mechanical advantage is 4. It will only take about 1/4 the weight of the object to pull it up _except_ for over-coming the <u>friction</u> of the lines going through the pulleys.

After explaining how to determine the mechanical advantage of a pulley system, use the reproducible sheet on page 15 for students to confirm what they know.

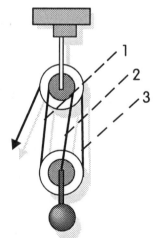

3 strands holding weight
M. A. = 3

Applications

Believe it or not, a mechanic can lift a car engine with two fingers using a pulley system with lots of pulleys in it. The more pulleys, the easier it gets.

Playground Physics

Pulleys

Date _____

Now that you have practiced using pulleys, use what you learned to answer these questions.

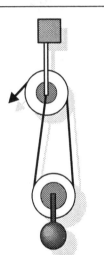

With the pulley on the right, it may be easier to lift the weight because you can hang on the line and use your own weight to help your muscles. On the other hand, a single pulley system doesn't provide a mechanical advantage that reduces the amount of force needed.

You can tell the mechanical advantage of a pulley system by counting how many strands of line are supporting the weight not counting the strand you pull on. What is the mechanical advantage here?

_____

This pulley system will make it easier to lift the weight. If you pull down 4" (10 cm) of line, how far will the weight go up?

_____

Because the weight only goes up half as far as you pulled down, the

mechanical advantage is _____. Write the numbers 1 and 2 on

the strands that are supporting the weight in this picture. Those are the

two lines that told you the mechanical advantage of this pulley.

This pulley system should make the lifting much easier.

What is the mechanical advantage here? _____

If the weight was 100 pounds or kilograms, how much force would be needed to pull on the line to lift the weight easily?

_____

If you wanted to lift this weight 8 feet (about 2 1/2 meters), how much line would you have to pull out?

_____

What do you call the scraping and rubbing that keeps machines from getting full mechanical advantage?

_____

Name _____

Simple Machine Combinations
Student Sheet

Date _____

We often use two or more machines in combination.
Label the two types of simple machines that you see in each of the pictures below:

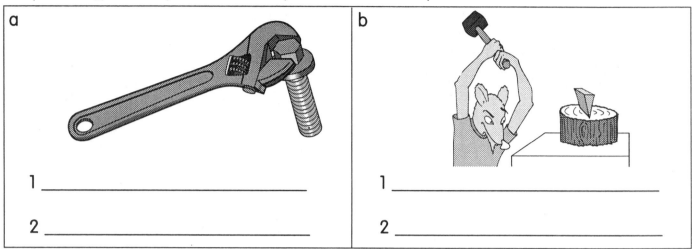

a

1 _____

2 _____

b

1 _____

2 _____

Some devices are themselves two kinds of simple machines put together in the same device.
Label the two types of machines that are combined in each of these devices:

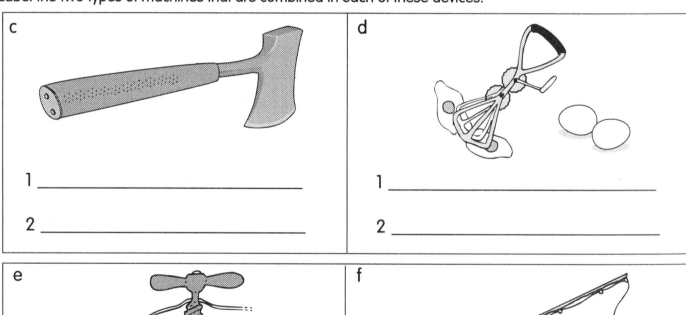

c

1 _____

2 _____

d

1 _____

2 _____

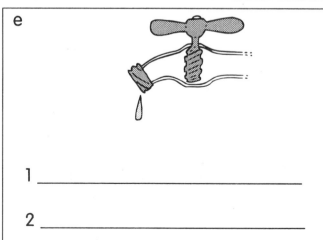

e

1 _____

2 _____

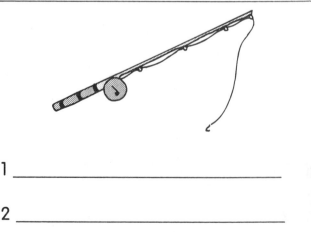

f

1 _____

2 _____

Playground Physics